Cotswold Privies

Uniform with this volume

EAST ANGLIAN PRIVIES

by Jean Turner

Cotswold Privies

by

Mollie Harris

photographs by

Sue Chapman

Countryside Books

Newbury · Berkshire

First expanded paperback edition 1995
First published in 1984
by Chatto & Windus

COUNTRYSIDE BOOKS
3 Catherine Road
Newbury, Berkshire

ISBN 1 85306 377 0

Produced through MRM Associates Ltd, Reading
Typeset by Acorn Bookwork, Salisbury
Printed by Woolnough Bookbinding Ltd, Irthlingborough

Sue and I would like to express our grateful thanks
to everyone who allowed us to photograph their privies
(unfortunately not all the pictures could be used),
and for the many anecdotes that we gathered from them.

Our special thanks go to Dr A. H. T. Robb-Smith,
Mr G. J. Gracey-Cox, and Oxfordshire Museum Services,
Woodstock, Oxon.
for their valuable information.

To *all* these people we dedicate
Cotswold Privies

Contents

This was our very first find in a neighbouring village.

FOREWORD

Sue Chapman and I travelled many miles gathering our collection of Cotswold Privies. Some of our journeys were very fruitful, while from others we drew complete blanks. Some people turned us down flat when we approached them, others gladly gave us permission to photograph their privies; some welcomed us, others were just downright rude.

Trying to keep secret our discovery of these long lost privies and the knowledge of their whereabouts has been difficult, to say the least. The reason for our near-secrecy was that we didn't want anyone else to 'do' Cotswold Privies before we had the chance to get our joint venture in print. So at first we told only a few trusted friends, living in a fairly scattered area. They then put out feelers for us, sometimes coming up with little gems. So it is to them that we have to be thankful, for without their help, and the co-operation of the folk who allowed us to photograph their privies, this book would never have got off the ground.

Sometimes the privies were still in use; sometimes the buildings were being used as coal stores, garden sheds and chicken houses; and sometimes they had been carefully preserved, as if they were waiting for people like us to come along and record their very existence, before they are thoughtlessly destroyed and gone for ever.

We often wondered why two- and three-seater privies were built, without partition between them. Did the family go 'down there' together, to sit and chat or plan the day ahead? Did the young brides and grooms actually use those straight-across two-holers, with one hole bigger than the other? Did madam sit on her small hole and the man of the house on the big one? We know why the two-holer, with the little one built on the end, was so designed – so that mother and child could sit there at the same time, and the mother could show the child what to

The author's joy at finding the three-holer at Kelmscott Manor.

We found this wonderful three-holer at Kelmscott Manor. The house is famous for its association with William Morris, father of the Arts and Crafts Movement, who rented it from 1871 to 1896. *Below left*, front view, *below right*, back view.

Most of our information came from chatting to villagers.

do – how to pull the paper off the nail at the back of the door, and so on. I did hear of a five-holer! Imagine all those people sitting together – or even wanting to go at the same time!

Cotswold Privies was intended to be just that – privies from the Cotswold area. When it became known that Sue and I were looking for these ancient relics we had phone calls galore, all of which we followed up. Several came from our native county of Oxfordshire – we even learned of one in our own village! Some real gems came from other parts of the county, and we have included them in our collection – with apologies to the Cotswolds.

So we present *Cotswold Privies*, Sue Chapman and I, hoping that it will bring back 'nostalgic' memories to the older generation and enlighten the younger generation as to what we old 'uns had to put up with years ago.

MOLLIE HARRIS

A long walk from the house in the rain, carrying the 'newspaper' with her.

Behind this door, made from old ship's timbers, was once a three-holer privy.

Mont Abbott shows us how the vaults or privies were emptied with a long-handled scoop.

[1]

Privies of Yesteryear

Before you read about the old privies of *my* youth, and the ones that we found in the Cotswolds and the surrounding countryside, I'd like to tell you a little of the history of the privy and the conditions that people – even royalty – had to put up with.

In mediaeval times plumbing, as we know it, didn't exist. It was the monks who took most trouble with this side of life. They sited monasteries with drainage in mind and wherever possible built their rere-dorters or sanitary wings close to a stream. In the castles the privies were called garderobes. Usually there was a sanitary tower with several privies to each floor, communicating with the moat by shafts. In other cases they were in turrets, with the privy projecting out over the moat into which all the contents of the privies fell. Sometimes they were built in the thickness of the chimney breasts, which kept them warm in winter, and the draught carried away some of the smell. Henry III was very particular about his sanitary arrangements, and when he was short of money and visiting the rich barons he always sent instructions as to the arrangements he expected.

The Chamberlain was responsible for supervising the sanitary arrangements in these great houses, and in John Russell's *Book of Nurture*, written in the fifteenth century, we read:

See the privihouse for easement be fayre soote and cleane
And that the bordes this uppon be covered with cloth fair and greene
And the whole himself look there no board be seen

Thereon a fair cushion the ordure no man to teme
Look there be blankit cotyn or lynyn to wipe the nether ende.

Lucinda Lambton tells us that public lavatories were often built over rivers and streams, and would have been used by the general public, because they would have been almost the only conveniences in the land. Only a few of the *very large* mansions and castles would have owned their own privies. So chamber pots were used in the houses overnight and the contents more often than not simply thrown out of the windows into the streets, often landing on the passer-by.

Where no running water was available great wells and cesspits had to be built, and the cleaning of these pits was so unpleasant that it was only carried out when the stench made it necessary. In 1281 the privy of Newgate Jail in London was emptied; it took thirteen men five nights to carry it out and cost £4. 7s. 8d., which was a considerable sum for the time.

But the privies and latrines, both public and private, had to be cleaned out now and then. And a cesspit with accumulated filth of months was a terrible job for the 'rakers' or 'gong-fermors' as they were called ... ('gong' from the Saxon *gang* to go off and *fey* from the Saxon word to cleanse) and their pay was quite good: forty shillings (£2) for each job. One such gongfermor, known as Richard the Raker, met with a dreadful death in 1326 when he fell through the rotten planks of his privy and drowned 'monstrously in his own excrement'.

Some of the old privies, as Dorothy Hartley says, had long seats with many holes, not only to accommodate many work people at once during a crowded dinner hour, but also to set aside a few holes over tubs for those who could 'oblige' and so collect the urine separately. This was used in some industries and provided a cheap form of ammonia.

Sometimes the richer classes minimised the smell by having

a pipe to connect the earth chamber, but in poor houses boards were set over a deep pit, and as the wood sometimes rotted there were many accidents caused by people falling through and drowning in their own filth. An account of such an accident appears in Fabyan's *Chronicles* (1516):

'In this year also fell that happe of the Jew at Tewkesbury which fell into a gong upon the Saturday, and would not for reverence of his Sabot day be plucked out, whereof hearing, the Earle of Gloucester that the Jew did so grete reverence to his Sabot day thought he would do as much unto his holyday which was Sunday and so kept him there till Monday at which season he was foundyn dead'.

In London in 1535 a raker was appointed to every ward, who went round three times a week sounding a horn, and everybody had to put his offal into the open street before five in the evening to be collected.

Care had to be taken when going to the lavatory, as a man called Boorde emphasized:

'Beware of draughty privys and of pyssynne in draughts, and permyt no common pyssyng place about the house – and let the common house of easement to be over *some water* or else elongated from the house.

Beware of emptyng pysse pottes, and pyssing in chymnes'.

This did not mean *down* or *up* 'chymnes', but pissing into the back of the fireplace, onto the accumulated wood ash.

As late as 1570 Tusser forbids 'pissing in chimneys', but some of the small closets, set into the wide stone chimneys (the sort of holes that get labelled 'priests' holes', 'bacon smokers', or just 'chimney-sweeping holes'), look very much like emergency privies, especially when they occur high up, on bedroom floors.

Even royalty had to put up with smelly cesspits, for Mary Queen of Scots wrote a letter to a friend during her imprisonment in Tutbury Castle in Staffordshire stating that:

'As no house, with so many low bred people in it as this, can be long kept clean, however orderly they may be, so this house, and I blush to have to say it, wanting proper conveniences for the necessity of nature, has a sickening stench ever lingering in it. On every Saturday too, the cesspools must be cleared out, even to the one below my windows, whence come none of the perfumes of Arabia.'

The ancient universities had some of the same problems. At New College, Oxford, the detached mediaeval building which housed the latrines still survives. Robert Plot mentioned the Long Room in his book, *The Natural History of Oxfordshire*, of 1677: 'I hoped it not improper to mention a structure called The Long House. I could not but note it as a stupendous piece of building, it being large and deep that it has never been emptied since the foundation of the college, which is about 300 years since, nor is it likely to want it.'

In fact the fifteenth-century Compustus Rolls do record the occasion of cleaning out of the Long Room in 1485.

In the seventeenth century Anthony à Wood wrote 'that the New College Long Room was built for a mean purpose'.

In 1880 the lower floor was expanded and earth closets installed. But early in the twentieth century the privies were destroyed and baths installed in the Long Room. During the Seventies, however, it was completely cleared and beautifully done up and now the room is used for concerts, dancing and meetings.

A trio of two-holers.

But a reminder of that period was saved and a Victorian urinal known as 'Jennys', which occupied one corner of the Long Room, is preserved in the Oxford City Museum.

At the end of the sixteenth century the first true water closet was devised by Sir John Harington, Queen Elizabeth's godson. He wrote an entire book on the subject. But the chamber pot was still the most popular convenience as this lovely poem, 'Piss-Pots Farewell', of 1697 shows:

> Presumptuous pisse-pot, how did'st thou offend?
> Compelling females on their hams to bend?
> To kings and queens *we* humbly bend the *knee*,
> But queens themselves are forced to stoop to thee.

Evidently Jonathan Swift despised the pot in the bedchamber or the dark, dank closet; and in his *Direction to Servants* of 1745, he writes for the house-maid:

'I am very much offended with those Ladies, who are so proud and lazy, that they will not be at the pains of stepping into the garden to pluck a rose,* but keep an odious implement, sometimes in the bedchamber itself, or at least in a dark closet adjoining, which they make use of to ease their worse necessities; and you are the usual carriers away of the pan, which maketh not only the chamber, but even their clothes offensive, to all who come near. Now, to cure them of the odious practice, let me advise you, on whom this office lieth, to convey away this utensil, that you will do it openly, down the great stairs, and in the presence of the footmen: and, if anybody knocketh, to open the street door, while you have the

* use the outside privy

Above, early water closet in blue and white china.
Below, chamberpot, kept under the bed.

vessel in your hands: this, if anything can, will make your lady take the pains of evacuating her person in the proper place, rather than expose her filthiness to all the men servants in the house.'

By the beginning of the eighteenth century water closets of the sluice type were being installed; Queen Anne had fitted at Windsor 'a little place with a seat of easement of marble with sluices of water to wash all down'. By then most of the larger houses had them installed. Often they were arranged with two seats so that one might have company, or were built in the garden in the form of a temple, like the one mentioned below.

Sir John Vanbrugh designed beautiful Blenheim Palace, Woodstock, Oxfordshire, the building of which was completed in 1715. While all the work was going on Sir John stayed at Hampden Manor, at Kidlington, whose owner supplied many farm carts, horses and men to haul the thousands of tons of stone that went to building Blenheim Palace.

Sir John was so grateful for his host's kindness to him during his stay that, in appreciation, tradition says, he told him that he would build him the finest lavatory in all England, and a water one to boot.

So he designed and built the castle-like lavatory (see page 27) over the main ditch drain, and instead of the 'soil' being retained in a pit, it was continually washed away. Apparently this very early ditch drain went through the village of Kidlington and on down Mill Street, Church Street and through the vicarage grounds and finally into the river Cherwell. In 1832, however, the vestry council decided that the ditch drain should be filled in. But the imposing castle-like building still stands in the grounds of Hampden Manor, Kidlington.

Towards the end of the century further progress was being

made in the improvement of the privy. The first patent for a water closet was taken out by Alexander Cummings in 1775, and that was followed three years later by a greatly improved valve closet devised by Joseph Bramah,* of hydraulic press fame. The nineteenth century saw many different inventions like the Rev. Henry Moule's earth closet and Parker's patent 'Woodstock' earth closet, photo on page 51 – and several types of water closets were patented. George Jennings was one very important water closet manufacturer at that period (see photo on page 72), but of course his 'Pedestal Vase' would only have been available to the wealthy folk.

At the International Exhibition of Hygiene at South Kensington in 1884 Messrs Doulton installed great urinals of earthenware at vantage points which the visitors might use free of charge, and there was a competition for the best water closet. There were thirty entries, and after considering shape and general comfort the judge submitted each article to a practical test, which, it was stated, he found very fatiguing. Into each pan were placed ten small potatoes, some sponge and four sheets of thin paper, and it was necessary that all should be removed at one flushing. Only three passed the test successfully.

These inventions were matched by an improvement in sanitary conditions in the slums. It was in 1848 that a public health act was passed making it law that a fixed sanitary arrangement of some kind, whether it be an ash closet, a privy

*So good were Joseph Bramah's water closets that the word 'bramah' was used to describe anything of first-rate quality.

As a family (during the Twenties and Thirties) we used the expression 'bramah' if anything was super – say someone had a new bike, or grew an outsized marrow, we would exclaim 'en 'e a bramah', none of us knowing where this expression came from until I started doing a bit of research on lavatories.

or vault type, a water closet or bucket, must be fitted to every household.

From that time the cesspits began to give way to a new and effective system of sewers. To Cloacina, goddess of the common sewer, one Victorian prayed with these delightful words:

> O Cloacina Goddess of this place,
> Look on thy servant with a smiling face
> Soft and cohesive let my offering flow
> Not rudely swift nor obstinantly slow.

Conditions didn't, however, change fast in many small towns and villages where primitive lavatories were still in use after the Second World War. Indeed in one remote village Sue and I found bucket lavatories still in use in 1984.

At Burford, twelve miles away from my home, comes a report of 1871 by Dr T.H. Cheatle, Medical Officer to Burford District, in which he stated that as most cottage gardens were very small, the privies were close to the houses and were nearly all planked vaults, which allowed free penetration into the surrounding soil. The privies were deplorably defective and in some instances there was only one to three or four houses, the privy, the pigsty and the well being within a few feet of each other. This report was made at the time when scarlet fever, diphtheria and consumption (TB) were rife, and no doubt such conditions were one of the prime causes of such diseases.

I end these snippets of history with a little gem from the parish records of Kirtlington in Oxfordshire. In the year 1872 a relieving officer applied to the bench magistrates for a summons against Mrs Ann Fathers of Kirtlington, Oxon, to show cause why she did not obey the orders of the Inspector of *Nuisances* in cleaning a privy which was so full that it was overflowing the seat and thereby causing a great nuisance.

We have certainly come a long way since then.

'The finest lavatory in all England', at Hampden Manor, Kidlington.

[2]

What did Folk use before Paper?

I've been trying to find out what people used before paper: I mean before newspaper.

At one time in the North of England sycamore leaves were used – but what in the world did they do in the autumn and wintertime, for sycamore leaves are among the first to fall! From Winchcombe in Gloucestershire came the story of the use of 'rags'. People used rags for centuries, and continued to do so if they were poor and couldn't afford a daily paper.

When we were children and were 'caught' up in the fields, we had what we called a 'country one' – in any old ditch or hedge, using a good handful of grass or a couple of dock leaves – and farm workers tell of using hay and straw for their ablutions. The Romans, who settled in many places in the Cotswolds – Cirencester, Northleach, in fact all along the Fosse Way – were a very clean race; they washed themselves afterwards, and used sponges on sticks which were kept in containers of salt water or dipped into running water. (There is a sad story in Seneca of a German who mistook the sponge for something to eat and died of it.) Poorer Romans might use a stone shell or a bunch of herbs to clean themselves. But the ladies liked to use a goose feather because of its softness.

In one lavatory that I went into some time during the Thirties there was a notice over the door, burnt out on wood, saying 'Yer tis then', and inside, pinned up beside the old privy, was a scribbled note which read

If in this place
You find no paper
Behind the door
You'll find a scraper.

Mr Abbott's old privy, and the sheets of newspaper.

Whether this was a joke or not I never found out. Thankfully there was a good wad of conveniently-sized pieces of newspaper, strung on a string which was hung on a nail behind the door, but the rhyme reminded me of a little jingle that we used to chant in the days of my extreme youth:

In days of old
When knights were bold
And paper wasn't invented
They used blades of grass
To wipe their arse
And went away contented.

Lavatory paper or toilet paper – toilet tissue as we know it today is a fairly recent innovation – has only become widely used since the Thirties. I had never used proper lavatory paper until I started work – before then it had always been newspaper.

Below is a photograph taken by Sue of what must have been one of the first toilet-paper holders ever made. I understand that it was made by Cameo Toilet fixtures, established in 1880. It was given to us by a lady who also let us take a photograph of her privy building.

Probably one of the very earliest toilet-roll holders.

[3]

THEM WAS THE DAYS

To combat the smell, sweet-smelling shrubs were grown near, up against and over the privies – honeysuckle, roses and lavender being the most prominent. Often a good big bush such as box would be grown in front of it, giving the person who entered a little privacy, and hiding the place from the view of the passer-by.

On the other hand, the privies serving the folk who lived in a row of cottages were usually built in a row at the top of the gardens, so everyone knew exactly when anyone else went along to relieve themselves and, to use one old boy's expression, 'you had to watch yer time'.

Others, and we came across a number of these, consisted of two privies in the same building – with a door at each end, sort of back to back, separated only by a partition, they served two different families. Here again, 'you had to watch yer time'. One old lady told us that she'd had one like this when she was young and first married and the man who lived next door, she said, 'did it on purpose, – no sooner than I'd got meself set down, unbeknown to me he'd creep into hissen and start a chattering all cheeky like. I had to tell my husband in the finish and he told him awf, you see I'd got frightened to go down there in case he come and set next door, 'cos you could hear every little sound, 'twas no good pretending as you wasn't in there.'

And her saying this reminded me of something that happened when we were young. Our 'two-holer' was side-by-side with missus next door's, and here again only a wooden partition separated us. Well, she used to sit on hers and sing. At this particular time the song 'I'm forever blowing bubbles' was

Ivy-covered privy in Oxfordshire.

all the rage, and one day she was sat in her privy trilling away, 'I be ferever blowin' bubbles', and my elder brother who was sat in ours shouted through to her, 'Ah that ent the only thing you be blowing missus.' Of course she told our mother, and as a punishment he had to help our stepfather dig the allotment all day on the following Sunday.

In my young days, our old lavatory was situated well away from the cottage up the top of the garden by the pigsty, for the same reason. Sat there on the throne it was a place of peace and quiet, with the morning sun shining warmly on your legs, which were placed in a most strategic position, with one foot either side of the half-open door. The left foot was used to keep the door at that angle, while the right one had a more important job. If someone came whistling up the dirt path from the cottage, to pay an urgent call, a quick flick of the right foot slammed the door shut, a method which was used by all the family.

The lavatory was a place to sit and dream and read. Often I read more from the newspaper which was hung on a nail behind the door than I ever did in the cottage. The only trouble was that by then the paper had been torn into a wad of conveniently-sized pieces, a hole had been pierced through the wad of paper with the aid of an old meat skewer, and the whole lot strung up together with string so that the user could snatch off a piece when it was needed. You would just get to an interesting bit of reading, only to find, after a frantic search through the remainder hung behind the door, that a previous 'visitor' had already made good use of it. Then it was back to last year's *Old Moore's Almanac* to find out what terrible things were supposed to have happened to us.

Comfortable and warm, time just stood still as you sat and watched spiders weaving big wonderful webs to catch the many flies that always seemed to be in there. Then suddenly the air might be shattered with an urgent cry from outside: 'Come on our Mollie, you bin in there half an hour and I wants to go ever so bad.' Still I'd take no notice – my sister might just get fed up and go away. Then more urgent cries: 'Quick hurry up, or I shall do it in me knickers.'

'Go on up Strange's field then and have a "country one",' I'd shout, 'I ent coming out yet fer you er nobody.' With cries of 'I shall tell our Mum' I'd hear her slink off.

With the door still shut in case she came back I'd sit there quite happily, the light from the sun filtering through the designs cut out of the wooden door (really for fresh air). We just had the shape of Vs cut out at the top and bottom of ours, but Mrs Bone further down the village had hearts cut in her lavatory door, *and* she had a religious text hung up inside as well. Lovely it was, a picture of a spray of apple blossom with the words 'HE WATCHETH OVER ME' painted in gold letters. I went in there once but never again. I didn't think it was right for me to expose my bottom with 'Him' watching.

Our lavatory at this particular time was a privy or vault type, and the sort that most everybody in the village had. It was a huge deep hole in the ground and over this was built a box-like contraption with two holes or places to sit. This was a 'two-holer', but some families had three- and four-holers. In ours was a large hole for grownups and a smaller one for children, which was set almost a foot lower than the grown-up one, so that the young child could sit on it with ease. It was rather a nice arrangement really – mother and child could sit there at the same time, and the child could learn what to do. The holes should have had wooden lids on when not in use, but not ours; they had most likely gone up the copper hole years ago to help boil the washing up.

Privet and other shrubs hide the building from the roadway.

In many country areas, especially the more remote villages and hamlets, earth closets or privies were still used up until the mid-Thirties. These only had to be emptied about once or twice a year depending on the size of the family using the privy.

Well, our privy was emptied about twice a year, when a man came with what we children called the 'lavender cart', a horse-drawn vehicle with a big steel barrel like a container on the back into which the contents of the privies were emptied. It had a sliding lid which should have been kept shut, but often the old fellow who drove it would ride along with the lid open, and a terrible smell seemed to linger all up the road long after he had taken the contents away to wherever he took them. He often came very early in the morning, and we were fascinated to see him ladling the contents of our vault out with what looked like a very big spoon on the end of a long pole.

On page 16 Mont Abbott, holding a thing that he calls a 'shittuss scoop', shows how this unpleasant task was done. In some villages the housewife would save her ashes and cinders from the fire and begin to make a stack of them near to where the earth closet or privy was situated. When the ashes were about three feet high, the man of the house would get this long-handled, bowl-like 'shittus scoop', and from a trap door at the back he would start to ladle out the contents of the privy and tip it on to the pile of cinders, which acted as a sort of filter. When it came round to gardening time the husband would shovel the cinders on to his garden and dig them in. And the wife would then start another pile.

After a while – and I can remember this happening to us at Ducklington – the earth closets were replaced with buckets. Quite a number of earth closets have been preserved, though they are not in use now, of course.

The buckets were housed in the same buildings where the

Mrs Clarridge's outdoor privy, still in use. We asked about the chair. 'That's for the ladies to put their coats on,' she replied.

privies had been: the big hole in the ground had been filled in and cemented over, and replaced with a bucket like the one on page 59. The same wooden seat was used but with a trap door, either at the back or front, so that the bucket could be taken out and emptied – mind you, some of the earth closets had a door at the back too, for emptying purposes.

The installing of the buckets was supposed to be a sign of progress, a nicer, more modern way of life. But my mother never thought that it was. We were a family of seven growing children, plus our parents, so the bucket was soon filled; and every night, when my step-father pushed his bike in the yard,

Left, this idyllic country scene would fool most folk, but note the corrugated iron nailed on the roof of the lavatory. The lady who lives in the nearby cottage has no electricity, her water supply comes from a spring in the bank just outside her cottage, and halfway up her garden is the bucket lavatory. Her garden was superb!
Above, inside view of the lavatory, still in use.

my mother was waiting for him, standing on the doorstep, arms folded across her chest, and she'd say, 'It's no good, it's full up and running over, and you'll have to go and empty it.' So before he had his tea, he had to go up to the top of the garden, dig a damned great hole and empty the contents of the bucket in it.

In quite a number of the places that Sue and I visited, buckets were still being used. No wonder those folk still grow super vegetables!

To the traveller in 1939 the quiet Oxfordshire village with its thatched cottages and roses round the door, and cattle in the fields, chewing the cud contentedly, knee-deep in wild flowers, must have seemed an idyllic spot. Yet there were no amenities in the village at all – people used bucket and earth closets, water was drawn from wells in the garden, and the cottages were lit with oil lamps and candles.

Then World War Two broke out, the powers that be decided that they would build an aerodrome on the outskirts of the village, and workmen were drafted from all over the country to get on with this vital job. Now a man and his wife were living in a little tumbledown old cottage in the village, and through the post they heard that their nephew from London was coming down to help build the aerodrome, so it was agreed that the young lad should lodge with his old uncle and aunt while he worked there.

'Twas a lovely sunny Sunday morning when the lad arrived, he'd never been out of London before in his life. He reached the cottage, there was his old uncle sitting under the apple tree where he'd been shelling the peas for dinner. The young lad went into the cottage to say hello to his aunt and then he came out a bit quick and said to his uncle, 'Uncle, where's your toilet?' so the old chap said, 'Wurs the what?' 'Well, where's your lavatory then?' the lad enquired a bit urgent like.

'Ah, thas the place, that little old stwun place, thur at the bottom of the garden, behind the box-bush.'

So the lad went down, but came back up the garden after about a minute and said, 'I can't go in there, uncle, 'tis full of flies.' The old chap took out his pocket-watch, glanced at the time and said, 'Ah thee try and hang on fer about ten minutes, missus is going tù dish up the dinner, and they'll all come up yer then.'

from A KIND OF MAGIC

After we had been living in our cottage for about seven years these awful vaults were replaced by bucket lavatories, but not before an incident happened that could have easily been a tragedy in our family. I must have been about nine or ten at the time and young Ben, one of my stepbrothers, about four. He had got into the habit of waiting to answer nature's call in the evening just as he was undressed ready for bed. He would say, 'Mum, I want to go to the lavatory,' and our mother would say, 'Mollie, take him down,' and I, no bigger than two penn'orth of 'apence, would stagger down the dirt path with Ben on my back, our light a candle in a jam-jar that often blew out on the way down.

Ben was a cheeky, spoilt showoff, and each night as I carried him down he would jog on my back and whine, 'I wants tu go on the big seat, I wants tu go on the big seat.' I suppose he

A two-holer in a quiet Oxfordshire village. The vault or soak-away was built in brick with fine arches. The building, covered in ivy and other climbers, was well away from the beautiful Tudor farmhouse.

thought he was old enough, but of course he was much too small.

As the weeks went by I got fed up with this nightly ritual, then one wet night I didn't bother to sit him carefully over the small hole but backed straight on to the big one and let him go. He folded up just like a shut-knife and went down the hole backside first, leaving only his head and hands showing, and it wanted emptying. He let out a blood-curdling scream and I yelled at the top of my voice – the noise brought the whole family running pell-mell down the garden path. They hauled Ben out, smelly and frightened. Ben had a good hot bath and I had a damned good hiding. But that 'larned' him. He answered nature's call earlier in the day after that.

[4]

Too Late

So many times Sue and I heard of a privy that might just still be in use in a remote hamlet way up in the hills, only to find when we tactfully enquired about it from the owner that it had been destroyed. Here's a typical reply:

'Till about two months ago we had one down the bottom of the garden, used it the best part of fifty years we did, then the council give us a grant to have a water closet built indoors, ah 'tis like being in Buckingham Palace to sit thur in the warm, arter all they cold freezing times we had tu set down thur in that old 'un', an old man said, pointing to a little ramshackle building thirty or forty yards away.

'Have you left any furniture in there?' I enquired.

'We never had no chairs ner nothing in ours,' he replied, looking a bit puzzled, 'We was only too glad to get out of thur pretty smartish I can tell 'e.'

'I mean the old wooden seat, is that still there?'

'Bless 'e no, we was that glad tu see the backside of it, I smashed it in and put it on me bonfire, but I still uses the little house, as we allus called it, but now I got a 'lectric pump in thur, to pump the tackle from our new lavatory down into the cess-pit, you see we got a water closet, but we still ent on the main sewer, none of us in the village be.'

We heard of a privy 'astride a stream' in the Cotswolds and set out on a forty-mile journey in search of it.

Alas, after several enquiries from the elderly locals, one man was able to tell us that it was 'all washed away' in 1976 when 'we had a tremendous lot of flooding hereabouts'. Then he went on to tell us about when he was a lad. Thirteen years old he was, and he had just started work at 'the Big House', a common expression used by the locals for the Lord of the Manor's abode.

One of his first jobs each morning was to collect all the ashes that had been taken up from the many fireplaces, and sift them. The very fine ash was tipped into a bucket, and 'then I took it down into a little wood, about two hundred yards from the mansion. In the middle of the wood was this lovely earth closet, you see the master didn't believe in the new-fangled water closets that was being invented. That privy was just like a little palace and as clean as a new pin.

'And now I'll tell you what that fine ash was used for. I used to tip it into a box that was kept on the right hand of the privy seat, ah, about a foot away from the hole. There was a little brass shovel there and what the master done was to sprinkle some of that fine ash down the hole, after he had used the privy, so there was very little smell.

'Course, when the old squire died, ah, about 1936 it was, and the place was sold, one of the first things that the new owner done was to have that lovely privy and the trees that surrounded it destroyed – I knows 'cos I was still employed there.'

[5]

THINGS THAT HAVE GONE DOWN THE PRIVY

Jack Smith's Auntie Lizzie was staying with Jack and his wife for a few days. Auntie Lizzie was never seen without her hat, a wondrous feathered creation that she had acquired from a jumble sale. It had belonged to Lady S., and Jack swore Auntie Lizzie wore the hat in bed. She even had it perched on her head when she went down to the old privy.

The folk next door also had a visitor, a young lad of about eleven years old, and he was proving to be a bit of a nuisance, because all he seemed to do was to take pot-shots with his catapult at everything that moved.

One day Auntie Lizzie was down in the old privy, 'doing her bounden duty', as she called it. She had just got up off the seat to adjust her clothing, when the boy, who was in the garden, spied through the fresh air cuts in the door what looked to him like a lovely cock pheasant in Jack Smith's privy. He took careful aim.

Suddenly all hell let loose. Out of the privy, screaming her head off, rushed Auntie, with her long white drawers round her ankles. She ran a few yards and then fell headlong on to the path.

''Tis gone down the 'ole, 'tis gone down the 'ole,' she bellowed, 'somebody get it out quick, 'tis gone down the 'ole.'

Hearing the commotion, Jack and his missus came tearing out of the cottage and ran down the garden path and picked the screaming old body up – then they realised that poor Auntie Lizzie was as bald as an egg.

And her lovely feathered hat was lost in the depths of the privy.

Spring in the Cotswolds. Note the slate-covered privy.

A Policeman's Lot

As told to me by an old fellow in a North Cotswold village.

'You might think what I be goin' tu tell 'e is a bit far-fetched, but it's gospel truth. You see it happened to me oldest brother who was the village bobby hereabouts...

'P.C. Pike. Percival Pike was his name, only the village boys called him Fishy Pike behind his back. Well, every morning he had to bike up to the nearby town to report to his Super, and

Front view of the same privy.

collect and receive any orders from him. On this particular morning he was, as usual, all smartened up, boots a-shining like glass bottles, uniform brushed and pressed, and his helmet on his head. He was just about to leave when he felt a very urgent need to go down to the privy. "Shan't be a minute," he called to his missus.

'After about ten minutes his missus began tu get a bit anxious like, so her sets off down the garden to see wur he'd got to. Well, he was just a-coming out of the privy door; his helmet had dropped down the hole and he had bin all that time a-fishing it out. He said afterwards that he had just bent over the privy to do number one as we calls having a pee (number two being the sitting sort) and straight down the hole his helmet went.

' "What shall I do?" he cried. "Look at it and smell the darned thing, and I ent got another one."

'Together in the kitchen they tried to wash off the muck, and time was a-ticking away. "I shall have to pedal like hell or I shall be late – and you knows how perticular the Super is about keeping good time."

'He arrived at the station hot and flustered and reported at once to the Superintendent. After a minute or two he said, "By God, Pike, have you stepped in something? There's a most terrible smell in here."

'Well, he thought as he'd better act the part, so he picks up one foot and then t'other, pretending to look underneath tu see if he had stepped in some dog's mess.

' "No sir," he replied, "I've not stepped in anything." Just at that moment a blowfly started a-buzzing round his helmet, followed by another and another and another. The Super noticed this and bellowed at old Perce, "It's you, you fool, it's you what stinks, go home and have a damned good bath, and for goodness sake get yourself clean, and report back here to me by one o'clock."

'Red-faced and sweating, poor old Perce left the room, but before he left the building he looked in at the clothing stores. There was nobody about in thur, so he grabs the first helmet he saw and went hell-for-leather home. But the helmet he grabbed was two sizes too small for him. It sat on his head like a pimple on a pie crust, so the village lads give him a new nickname. "Pimply Pike" they called him after that. Oh, I often wonders what they would have called him if they'd met him on the morning when his helmet fell in the privy.'

The ash closet with lever which released ashes to cover the contents of the closet.

Bert had accidentally dropped his jacket down the earth privy and was poking about trying to retrieve it. His mate called out to him – he wanted Bert to help him move a chicken house. 'Whatever be you doing in thur? You bin in thur this ten minutes as I knows on.'

Bert said, 'Well, I've dropped me jacket down the privy and I bin trying to hook it out.'

'What's the good of that?' his mate replied, 'you wunt be able to wear it again after it's bin down there.'

'No, I know I shan't,' Bert cried, 'but my bit of bread and cheese is in the pocket, thas what I be after.'

Clara's Auntie May from London had come down to the village to stay with Clara and her husband for a few days.

'Where's your loo?' she enquired after she had downed three cups of tea.

'Thur 'tis,' Clara told her, 'down the garden.' Off went Auntie May, and when she came back she looked quite worried.

'Well, that is a dreadful place,' she cried, 'and you got no lock on the door neither.'

'We don't need no lock on thur,' Clara's husband said, 'we bin livin' yer fer this thirty years and we ent never had a bucket of s--- pinched yet.'

Parker's Patent 'Woodstock' Earth Closet. This was an improvement on Moule's closet since it was fully automatic. When the user rose, the removal of pressure-activated levers released the earth or ashes from a hopper into a bucket. John Parker, a Woodstock cabinet-maker, patented his improvement in 1870.

[6]

TALES FROM THE COTSWOLDS

WINCHCOMBE

'When I was courting the young lady who became my wife,' one elderly man told me, 'she and I were both seventeen at the time. Her grandfather was ill, so we called one day to see him. The old fellow was propped up in bed.

' "Good morning Mr Arthers," I said, "And how are you today, sir?"

' "Well, young man, I be fairish today, I've got trouble with me bowels, you see — one day you could prop a five-bar gate up with it, and the next 'tis like water down a gutter in a thunderstorm." '

EYNSHAM

This happened to a neighbour of mine, now ninety, when he was young.

'I was sitting on the big seat of our two-holer doing what I had to do, course I'd got the door open, you never reckoned to sit in there with him shut. Well, 'twas a terrible windy day and as I dropped the paper down the big hole, the wind caught hold of it and it came flying out the little hole and landed on me head.'

Lady Cunliffe's beautifully restored privy.

SHERBORNE

One old fellow told me this story that happened to his cousin, Fred Fathers, and his wife Flo.

Fred had worked on the farm all his life, but a bad accident to his foot had finished him as far as farm-work was concerned. So his boss found him and his wife some temporary work as caretakers of a big old empty house; it was to be put on the market, and Fred and his wife were to show prospective buyers round the place. Well, the first 'customer' was a portly gentleman who seemed quite interested in the place. After he'd had a good look round he turned to Fred and said, 'How many windows face north, and where's the W.C.?'

Course Fred didn't know what he meant, so he looked at his wife and she whispered to him 'Wesleyan Chapel, you fool.'

Then Fred turned to the gentleman and said, very important like, 'Thur are five windows that face north, sir, and the nearest W.C. is half a mile away. I went six months ago, but I had to stand up all night so I ent bothered since.'

RISSINGTON

During the summertime our mother used to hang sprays of elderflowers up in our old privy. The smell from the leaves and flowers was supposed to keep the flies away.

Mr Hayward thought he had a three-holer but could only find two.

Mr Hayward's privy now assembled.

CHIPPING CAMPDEN

As told by George Hart (Jethro Larkin of 'The Archers')

Frank Jenkins had been constipated for nearly a week, so he went along to the chemist to ask if he could give him something that would relieve him.

The chemist, a friendly sort of chap, said that he had just the thing, but it had to be measured out very carefully and taken last thing at night. Taking a jar of white powder in one hand and a small empty container in the other he said to Frank, 'How far from the cottage is your lavatory?'

'Oh, about thirty-five yards.' So the chemist shook some powder in the container.

'Have you got a bolt on your back door?' he asked. Frank said he had, and the chemist shook a little more powder in.

'How many steps up your stairs?'

'Ah let me think,' Frank said, 'we got ten, yes thas it, we got ten of um.' And the chemist shook a little more powder in.

'Do you wear a belt or braces on your trousers?'

'Both,' Frank replied, and the chemist shook more powder into the container.

'There you are,' he said, 'that'll do the trick, but as soon as you get the urge you must go at once. That'll be sixpence,' he added.

A couple of days afterwards they met in the street. 'Well, Frank,' the chemist asked, 'did my medicine work all right?'

'Oh, ah,' Frank replied, 'it worked all right, but you beat I by three yards!'

HINKSEY

Another old fellow we met on our journeying told us how he always dug a hole in his garden in the daytime, but left it till it got dark before he 'mixed the puddin'' (as he called burying the contents of his bucket lavatory).

One frosty night he was doing just that when two fellows walked by.

'By goy,' one of them said, 'what's that terrible smell?'

'Tis only old Bert mixing his puddin',' the other laughingly replied.

To use old Bert's words, 'that always do smell summut dreadful on a frosty night, but by goy we did grow some wonderful crops on the bit of ground, and the rhubarb was most flavoursome. My missus used to make wine from some of it, and I've never tasted wine like it afore or since.'

My Auntie Flo

As told by Bob Arnold (Tom Forrest of 'The Archers')

'My Auntie Flo was twenty stone if her was a pound, a short rounded woman and nearly as broad as her was high.

'Their old vault lavatory had recently been done away with, and a bucket type put in its place. Theirs was the sort with a hinged wooden seat that you lifted up, from the front, when you wanted to take the bucket out for emptying. Course some of the folks who had bucket type privies had to take the bucket out from a trap door at the back.

'Anyhow, unbeknown to my Auntie Flo, her husband, my Uncle George, had emptied theirs before he went off to work on this particular day.

'About mid-morning, down the garden goes Auntie Flo and backs into the lavatory – her was so fat, and the building so narrow, 'twas only just wide enough for the bucket, so that her allus had to back in. Down her sets and then there was an almighty scream. Uncle George had forgot to put the seat back down, and poor old Auntie Flo was wedged fast in the bucket and her couldn't get out. And the poor old gel had to set there for nearly four hours before somebody heard her shouting and a-screaming.

'That took two strong men, Trueful Smith and Jimmy Pearce, over half an hour to move her out of that little building, pritnear dragging her arms and theirs out of their sockets in the bargain. Her backside was wedged that tight in the bucket, they said 'twas like trying to take a bung out of a five-gallon beer barrel.'

Lavatory- or privy-bucket. Note the short handle but long bucket to prevent the contents from spilling.

'Lèse-Majesté' by E.Y. Harburg

No matter how high or great the throne
What sits on it is the same as your own.

BROADWAY

Every time the man who emptied our lavatory bucket came, my mother used to give him what she called 'a cup of dutch courage'. This was a cup of whatever home-made wine that she had on the go. 'Twas an awful job, she thought, for old Jossop, having to empty other folks buckets.

Well, it happened to be a few days before Christmas, so as a bit of a treat my mother poured a generous measure of whisky into the four-year-old dandelion wine. 'There,' she said, 'that'll warm the cockles of your heart.' But it did more than that. We watched him from the window as he walked almost legless towards the back of our lavatory. With difficulty he opened the trap door and dragged the full bucket out. Just at that moment he seemed to lose his balance and pitched forward, landing head first in the bucket. And he, the bucket and the contents rolled over on the path.

We rushed out to help him to his feet. 'I'll fetch a bucket of water,' my mother cried, and the unfortunate fellow stood there while she chucked bucket after bucket of cold water over him.

'Where's me dawdy,* where's me dawdy?' he kept saying over and over again.

'Where's your what?' my mother cried.

*Dawdy – country name for trilby hat.

The tree and wheat are growing on the roof of this one-holer in a quiet hamlet. Like many others that we found, this had once been a vault type but was converted to a 'bucket' later.

'Me dawdy, me hat,' he said. Suddenly he leaned forward, 'Yer 'tis,' he cried, picking up a very wet and mucky object off the path.

We had no clothes to offer him, so he staggered off home to change, leaving his horse and 'lavender cart' outside our house. He arrived back about an hour later looking rather uncomfortable in a navy serge suit that smelt strongly of moth-balls, and wearing a very new-looking cap.

' 'Tis all I got,' he said, 'The missus is going to wash me working clothes today, I hopes her gets 'um dry quick, I do feel so silly in this yer get-up.'

We heard later that because it was nearly Christmas old Jossop had been given several drinks of wine on his rounds that day. Evidently my mother's drop of 'dutch courage' had been the last straw.

Here are some of the other little stories we heard as we went round the countryside looking for privies:

Memories of a Land-Army Girl

'There were six of us girls working on this big farm during the war and paper was that short, so the farmer's wife gave us a big old Bible and we used the pages out of that. It lasted us quite a long time and we were sorry when we'd used it all, it was such lovely soft fine paper.'

AS TOLD BY JENNIE FROM STOW

'My mother told me this story, and it was supposed to have happened to her grampy, who had two wooden legs. In them days they was just straight stumps. Well, the old fellow went down to the privy last thing at night before going to bed, and he was sat there with his two wooden legs thrust out in front of him. Now the old man's son had been out at the pub all night and was pretty drunk. Anyhow, before he went indoors he thought he'd go to the lavatory. In the gloom and in his drunken state he walked into the ever-open doorway of the privy and cried, "Somebody's left the bloody wheelbarrow in yer" and promptly tipped his father on the floor.'

KINGHAM

'My cousin was staying with his very old aunt in the country. Hers was a bucket lavatory, and the man next door always emptied it for her. There was a trap door at the back of the lavatory, so he would just open the little door, take out the bucket, bury the contents in the garden and put the bucket back again. Well, my cousin didn't know nothing about this arrangement and he was sitting one day doing what he had to do when he felt a bit of a draught. He looked down and the bucket was gone. He said, "I didn't know what to do, whether to carry on and do it on the floor or what. As I sat there contemplating, suddenly the bucket was shoved back underneath me backside, so I just carried on." '

Above, a two-holer.
Below, the trap door is for emptying purposes.

In most cases the country dweller who owned a bucket lavatory emptied his own, sometimes digging a trench across his garden and gradually filling the trench up; soil would be thrown over it, and eventually the whole garden would be used in this way. Then next year's vegetables grown on that ground. But in the towns the contents of the buckets had to be collected. One woman who originally came from a large town 'up north', but now lives in the Cotswolds, spoke of the 'midden men' who collected theirs.

'Two men used to collect up our street, they had a sort of wheelbarrow thing that held a round container. Early every Tuesday morning everybody stood their buckets out on the front doorstep to be collected by the midden men. Some folks sprinkled ashes on the top of the full buckets – others put sawdust or mortar, depending on what they had got. Sometimes the midden men were late, and I've known the time when those full buckets have still been out there at dinnertime.

'Then one man wearing a yoke across his shoulders would fix a full bucket on each hook and then run to the wheelbarrow – the other man tipped the container slightly, so that the man with the buckets could tip the contents into it. By the time they got to the end of the road the container was full up, so they took it away somewhere, but not before they had sloshed some of the contents all over the road. My mother used to rush out and swill our step down with a bucket of water, and sprinkle ashes on the pavement by the house.'

FROM WINCHCOMBE

A Cotswold expression: 'If there's a terrible smell around ... 'tis as strong as a privy with the lid off.'

'Very handy that bath is. I uses it every Saturday night,' the lady of the house told me. 'Has me bath in front of the fire, 'cos thur ent nobody here but me.'

A bucket lavatory, converted from an earth closet. Little slots on the lid make it easy to lift. This very clean lavatory is still in use.

This is an old lavender cart, previously horse-drawn, with bigger wheels. Michael Harvey, employee of the Witney Council, is emptying a bucket belonging to one of the showmen at Witney Feast, always held in September. The towing bar has been changed but the old tipping mechanism is still in use.

[7]

LOCAL TALES

The year was 1947, it was wintertime, and freezing cold. The houses that were situated on a steep hill going out of the town of Witney still used bucket lavatories.* Every Tuesday night the buckets had to be brought through the cottages and put out on the side of the path, so that the 'lavender cart' man, Patsy Souch, could empty them early next morning. Most of the cottagers, not wanting to offend late night pedestrians who had to pass the line of buckets on their way home, sprinkled cinders and ashes from the fire over the top of the full buckets, so they didn't look or smell too bad.

On this particular very cold night, with the roads like glass, a heavy lorry got stuck halfway up the hill; the driver tried again and again to get it moving, but to no avail. Then he spied a bucket of ashes, in fact a whole row of them – probably put out to help unfortunate drivers up this hazardous hill on such a freezing night, he thought.

He grabbed the first bucket and sloshed the contents under his wheels ... needless to say, he was still stuck there hours later.

SNOWDROPS TO LIGHT THE WAY

A very well-known M.P. told me that in the village where he lived as a boy, everybody planted snowdrops either side of the path leading to the privy. On dark winter nights the heads of the white flowers helped to show the way to the thunder box.

*In 1952 there were still 250 households in the town of Witney that used bucket lavatories

THE FIREMAN'S STORY

''Twas 1953 and me and my pal hadn't been in the fire service very long. One night there was a call to go to the local filter beds, where the rubbish was blazing and getting out of control.

'The fire engine drew up fairly near the blaze. I jumped off and caught hold of the hose – my mate was a few yards behind helping to keep it well under our arms. Just in front of where we had been instructed to run the hose out there was a flat grassy area. We started to run over this towards the fire when I began to sink into something. I shouted "Quick, quick, help me, I'm sinking."

'I turned round and my mate too was fast disappearing. "Hang on to the hose," our officer called; somebody threw an old iron bedstead for us to hang on to, but that disappeared very quickly. I was now up to my neck and holding my head back to stop the muck going into my mouth. Suddenly we were both dragged back towards the fire engine – our mates were winding the hose in at breakneck speed, and very soon we were on firm ground again. By God, didn't we stink, so before they tackled the fire our mates hosed us down to try and get rid of the thick black jelly-like stuff that we were covered in.

'What we had run into was a sort of pond the sludge drained into after the sewer contents of the nearby town had been processed.

'The refuse lorries from the town were supposed to be emptied over this, but there was quite a big area that was exposed. We didn't suffer any ill effects from our near drowning, but it was an awful job to get rid of the smell.

'And I did hear that once, up at the filter beds, one of the 'lavender carts' ended up in the big pit-like place where they tipped their contents. What happened was that a driver backed his horse and cart just a bit too near the edge and the wheels began to slip. Quick as a flash he cut the horse free and stood and watched his 'lavender cart' slip into the pit, where it was swallowed up in the town's muck.'

'The snow was very deep on the ground,' one woman told us, 'and we had to cross the garden to the privy, crossing over a small drain that was sunk in the middle of the yard.

'Then our Dad thought of a good idea. He rigged up a rope from the back door to the loo, and dressed in coat and wellies we took turns to cross holding onto the rope, hand over hand, just like a trip to the Alps.'

In one village that we visited there was an old couple whose only sanitation was a thunder-box at the bottom of the garden. The railway ran along their bottom fence and it had become necessary to lay an additional line. For this it was vital to purchase a strip of land at the bottom of the old folks' garden.

The railway people offered to build a new W.C. adjoining the house. In due course this was ready for use. The old lady was the first to try it, but when after half an hour she had not come out her husband got worried. He banged on the door and called out 'Are you all right dear?' 'Well, I don't know,' came the reply. 'Every time I get hold of the chain to pull myself up there's a rush of water goes right through me.'

George Jenning's Patent Lavatory, mahogany-panelled, at Rousham House, Rousham Park. *Below*, a closer view of the hand-flush.

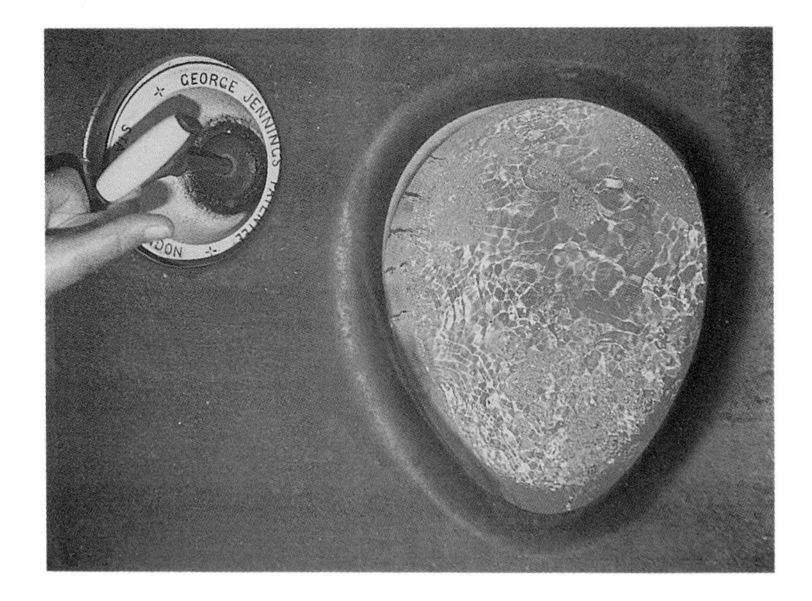

... fancy using that!

Astride a stream – a picturesque privy situated over a mill stream. (Author's coll.)

[8]

PRIVIES GALORE

A FONDNESS FOR FRESH AIR

It's funny, how some folk these days screw up their noses when earth closets are mentioned. Why is it, then, that they are remembered so vividly and with a sort of affection? And think of all the humorous stories that have been remembered, the accidents and happenings in connection with the privies of yesteryear. And how lovingly David Green wrote about his earth closet, in his book *Country Neighbours*: 'No indoor lavatory meant, amongst other things, a blow of fresh air, whatever the weather. In summer it had the attractions of a gardener's bothy, our lilac-shaded privy, and it made a capital place for musing on the garden's crops: or, with its view of bird-box and the birds that foraged among the beansticks (including a lesser-spotted woodpecker), was it to be scorned as an ornithologist's hide. There was also the wren that nested behind the door. There were drawbacks, of course. The privy might get snowed up entirely overnight, or a passing soldier in the weighty accoutrements of battle ('readin' and smokin' and makin' a weddin' of it', as a commiserator guessed) might, and in fact did, break the seat clean in two, but neither mishap was irremediable, and although I did the spadework myself I decided, at least when the damson was in bloom and the mistle-thrush singing in the orchard, that I wouldn't change places with anyone, no not with the most sumptuously porcelained and mahoganied millionaire.'

At the farm where I was brought up, we had a two-seater privy which was about twenty yards up the garden path. It had one large hole for adults and another about half the size which I and

An elderly lady still uses this well preserved two-holer at Leighterton, apparently by royal appointment. (Author's coll.)

my sisters and four brothers used. Like many other farms we had a pond, which was close to the privy. At a convenient height there were cavities in the privy wall and when the pond became full it over-flowed through the cavities and took much of the contents of the privy with it, and then it went into a ditch – and finally out into a stream. I have no recollection of it ever receiving any other attention.

Of course, city folk, at least during this century, didn't know much about earth privies. There was this young man, who had lived in London all his life and was used to water lavatories.

Well, he went 'up north' to visit an old uncle who lived in rather a remote hamlet. The young man eventually asked where the lavatory was.

'There 'tis,' his uncle said, pointing to a wooden building down at the bottom of the garden.

When the young man returned to the cottage he said to his uncle, 'That's a funny water system you got there. When I pulled the rope above the lav, no water came out.'

'We got no water down there,' his uncle roared, 'It's an earth closet, you bloody fool – you've let the pigeons out.'

A Miscellany of Lavatorial Items

Privies have given rise to some strange scenes and some strange ideas. Here's just a few of them, the first from a Mr Penny:

'I well remember my late father who came from the village of Aldbourne, Wiltshire), telling me that on leaving the village school he was apprenticed as a carpenter locally. He was taken one day to a remote cottage to assist an older carpenter in making and fitting a new wooden seat to the existing outside privy. The elderly woman occupant of the cottage was very noisy and kept interfering and getting in the way. The old carpenter decided to have a bit of fun with her and having made the new wooden seat and put it in place, he called the woman out to the privy and asked her to sit down on it. She asked why. The carpenter told her that as he had to cut out the hole in the privy top he needed her to sit on it so that he could mark round her backside in order to get the hole the right size!

That got rid of her – she shot off like a scalded cat!'

A rare inside privy found on a Gloucestershire farm. (Author's coll.)

A venerable brick-built example with a few loose slates, but otherwise serviceable. (Author's coll.)

It seems that, probably to relieve the tensions of war, servicemen in particular seemed to have derived a great delight in 'warming up' their mates while they were sitting on the communal latrines. Most, it seems, were built in row, with a channel of continuous running water serving the row of privies. The standing joke was to drop a piece of rag, well soaked in petrol, which could be lighted at the top end, the flowing water would gradually take it down the row causing pain and surprise to those sitting on those holes. One man who experienced this, said: 'I was at the top end, so it wasn't too bad, but them bottom ones copped it. Their privates was singed of hair as bare as a yew stick.'

This lovely memory came from a lady who lives at Shabbington, Aylesbury: 'I was born in a thatched cottage at Draycott, on the road from Ickford to Tiddington in Oxfordshire. We had a privy at the bottom of the garden and my sister and I used to cut newspaper in squares, which we all used in there: it was the only means we had – no such thing as 'toilet paper' in those days. At that time my uncle and aunt were in service at Waddesdon Manor, in Buckinghamshire, home of the Rothschilds. My uncle was head groom and my aunt cooked for all the staff who worked in the gardens.

Miss Rothschild used to give my aunt books that she had finished with – the *Tatler*, and things like that. My aunt passed them on to my mother when we called to see them. The magazines were not much good as 'bum fodder', much too stiff and shiny for that, you see. So my mother used to leave them in the privy for my dad to look at when he paid a visit down there. In those days, most everybody used to read in their lavatories.

One day I was sitting on the privy and thought I'd take a

look at the magazines, and out of one fell two one-pound notes!

Very excitedly I ran back to our cottage, my knickers half-up, half-down. My mother thought it was a fortune – it was more money than my father earned in two weeks. The only thing we could think was that Miss Rothschild used the £1 notes as *Bookmarks*!

When my dad came home we told him, and he ran down to the privy and went through every page of those magazines, but he didn't find any more.'

Did you know that George II died on the throne – literally?

He was in his closet when his valet, hearing a noise louder than the 'Royal Wind', rushed in to find that Death had levelled the king flat on his face on the floor of his loo.

From Queen's College, Oxford, comes in this item of news. In the year 1374, John Wyclif, the church reformer, who at that time was known as 'The Morning Star of the Reformation', apparently had his own privy to which he held the key. Quite a privilege as the other Fellows and students had to use a communal one. And in 1987 there seemed to be a great interest in the Oxford Union Debating Chamber. The powers that be wanted to change the 'Gents' Victorian bathroom by lowering the ceiling, so that a 'Ladies' toilet could be built in the 'air space'. Students said that if this was carried out it would ruin the bathroom's historical value: they pointed out that at least five Prime Ministers had peed in there – Gladstone, Asquith, Lloyd George, Macmillan and Heath had all at some time passed through the wooden door.

Mrs Rachell Hayes, of Birmingham Road, Redditch, discovered my interest in privies in unusual places and sent me some photographs of Abberley church clock tower. It was built in 1880, and right inside the tower is a privy – I wonder for whose benefit it was positioned in the church, the parson's or the congregation's?

A soldier returned from the second world war to find that the old earth privy at his home was still being used. He was all for building a lean-to nearer the house, and suggested that they should have an Elsan which would be an improvement on the old earth closet down the garden. This they did, but his father still liked to use the old one. So the soldier thought he'd soon make him change his mind, and watched for the old fellow to 'pay a visit', then he got a hand grenade that he had brought home. He took out the pin and aimed it near – not directly at – the old privy, the roof of which promptly blew off. Out came the old fellow, trousers around his ankles, and said, 'Good job I didn't let that fart awf indoors.'

A collector of period furniture was interested in an old chair:

'It's Queen Anne,' the salesman said.

'Oh yes,' said the collector, 'how can you tell?'

'Well,' the salesman replied, 'look at the letters carved on it: Q.A. That stands for Queen Anne, surely.'

'If that stands for Queen Anne,' cried the collector, not impressed, 'I've got a door at home that dates back to William the Conqueror.'

In the early days of motoring, a man was out driving his car. He passed through a village where there was a single petrol pump, but didn't stop as he thought that he had enough fuel in his tank to get home – a matter of about four miles. About a mile and a half along the road the car came to a halt, no petrol. So the driver looked round the car to see if he could find something to carry a drop of petrol – just enough to take him back to the petrol pump. All he could find was a small enamelled potty which belonged to his little girl.

He arrived back at the car and was just pouring the petrol into his tank when another car pulled up. The driver, a vicar, poked his head out, and said, 'I don't know what your religion is, but I do admire your faith.'

One day a man was out collecting rents on a council estate. Apparently in each building the back door to the houses and the one to the outside WC were side by side and quite close to one another. The rent collector, as usual, knocked, pushed open the door and called out.

'I've come for me dues, Mrs Jones,' he said, but had opened the wrong door, only to find Mrs Jones enthroned on her WC.

When there was talk of water being installed in a Cotswold village, one old lady was heard to remark, 'I shall stick to my old earth closet, that damp water in them newfangled 'uns would give I the piles quicker than anything.'

From the village of Leigh comes this shot of a privy christened by the family in residence as The End. It has a stone chute at the rear where the privy emptied itself into the remains of an old moat. (Author's coll.)

An English lady was to stay in a small German village. Not knowing any German she secured the help of a German schoolmaster, and wrote to him for information. One of her questions was to ask, was there a WC attached to her lodgings? The schoolmaster, not familiar with the abbreviation, thought that WC might mean Wold Chapel – which means 'Chapel in the Woods'. Thinking that she must be a devout Church-goer, he wrote her the following letter:

'The WC is situated some seven miles from your lodgings, in the midst of beautiful scenery, and is open Tuesdays, Thursdays, Fridays and Sundays. This is unfortunate if you are in the habit of going frequently, but you will be interested to know that some people take their lunch with them, and make a day of it, whilst others go by car and arrive just in time.

As there are many visitors in the summer, I do advise you go early. The accommodation is good, and there are about 60 seats, but if any time you should be late arriving, there is plenty of standing room. The bell is rung 10 minutes before the WC is open. I advise you to visit on a Friday as there is an organ recital on that day.

I should be delighted to secure a seat for you and be the first to take you there. My wife and I have not been for six months and it pains us very much – but it is a long way to go.

Hoping this information will be of some use to you.

Yours sincerely...

Living a I did in a small village in the 1920s meant that I was brought up in a world of privies, earth closets, or vaults, as some folk called them. Everybody in the village had one. Well, some of the big houses had two, one for 'them' and one for the servants and gardeners.

The privies of my youth ranged from one-holers, two-holers, three-ers and four-ers, and when I first started to make enquiries in the village where I live now, an old fellow said to me, 'You should 'a writ that book fifty years ago, thur used to be a five-oler in ower village.' A five-holer! Can you imagine that five people from one family would all want to go at the same time? I mean nowadays, even if you have a big family you are lucky if you have two lavatories, one up and one down.

These days we shut ourselves away and get on with our ablutions in private. But when we were young we often used to go down to our two-holer with brothers and sisters, and even the next-door neighbour's children. And it wasn't always just for 'business'. Oh, no! You lingered there for ages, to chat and giggle perhaps about a local boy who you fancied. And the smell that was always prevalent never seemed to bother us at all. We also found sanctuary down there, an escape from washing-up or taking younger brothers for walks out of our mother's way. On the other hand, if you could sneak down there on your own, it was lovely, away from the crowded cottage. You could just sit there, with the door three parts open, and the sun shining on your legs – to read from the squares of newspaper (the 'bum fodder' as we called it), or watch the birds in the hedges that separated us from one of the other cottages. There you could dream or have a quiet cry in secret: it was a peaceful place, until a shout from one of the family who was desperate to use the privy shattered your peace!

And I vividly remember, when we were young, we had oranges *only* at Christmas time. Two each, always, in our stockings. And in those days the oranges were wrapped in fine orange sort of tissue paper; we used to smooth it out carefully and use it when we went to the old privy. For two days we were in heaven! It was such a luxury, and a nice change from newspaper or *Old Moore's Almanac.*

Our privy also offered sanctuary for the spiders who wove huge magical webs down there. One old lady from the village used to come and ask my mother if she had any black cobwebs – good for stemming the blood, she reckoned. She never asked the missus-next-door for any: no cobweb ever had the chance to survive in her privy. My mother and the old lady would toddle off in search of these old black cobwebs (they had to be old ones) either in the privy or in the old wash-house which was a few yards from the cottage, and the old woman would go off pleased as punch with half a dozen cobwebs wrapped up in a bit of old newspaper. Of course the spiders lived there because of the flies and the flies lived there because of the smells and the rich pickings. I can't remember ever going down there – at least during the warm weather – and not finding a dozen or more 'blue-assed flies' buzzing about in there, especially when the six-monthly visit of the lavendar cart was due: our privy, with nine people using it, got a bit on the full side.

Some of the village privies were spotless. 'You could eat your dinner off some of the seats, they be that clean,' missus-next-door said one day. 'Who the devil would want to?' my mother cried. You see, missus-next-door only had one child, and our mother had seven, so you couldn't expect ours to be like Crystal Palace, even if hers was.

It was well into the thirties before some of the village privies, including ours, were changed to buckets – which was supposed to be a 'leap forward in hygiene'. 'Hygiene, my foot,' our mother retorted when a local councillor said this to her. 'You don't know what you're talking about,' she went on, 'there's nine of us using that so-called hygienic bucket – so what happens with my big, healthy, growing family visiting the privy at least twice a day, the damned bucket's full and running over by night-time, so it has to be emptied and the contents buried. The only blessing,' she went on, 'is that we shan't have to bother about manuring the garden, it'll be done for us.'

And t'was true, we and other villagers produced what Old Fred used to call 'jynourmus' vegetables.

A 'straight drop' privy and stone channel at Broughton Castle. (Author's coll.)

Marcus Binney, writing in *Country Life* in January 1985, about a great collection of stool-type conveniences amassed by Carl Gustav Wrangel in the seventeenth century, reported that one of them had a nice piece of verse inscribed in it:

Though your name be Modesty
Sweetly budding rose
Needs must when you enter here
Your private parts expose

There's an epigraph on a tombstone to an old woman buried somewhere in Gloucestershire – well, so an old Gloster-man told me – that reads so:

Where ever you be
Let the wind go free
For the queeze of a fart
Was the death of me

Anyhow, that would not have been good advice for the following:

According to the ancient Laws of Manu, every conceivable part of the body was liable to be amputated as a punishment, including the anus, of any citizen who might break wind in front of the king.
(Guido Manjno, *The Healing Hand: Man and Wound in the Ancient World*)

An old fellow in my village was asked why he spent such a long time in the privy. His reply was:

Well, sometimes I just sets and thinks
And sometimes I just sets and stinks
Other times I just sets

FEEDING THROUGH THE LEAF, WITH URINE

This delightful rhyme was sent to me some years ago and was evidently published by the *Northampton Chronicle* around 1930.

It is my firm belief
That feeding through the leaf
Will make all crops as healthy as can be;
And after careful test,
I find urine is the best ––
It feeds the plants and keeps them insect-free.

All plants do truly need
A much-diluted feed;
And this is how dilution should be done –
To eight pints of water
Add urine one quarter;
In other words, just thirty-two to one.

Sprayed gently on the leaf,
Above and underneath
It kills the pests and checks the mildew, too.
The growth it seems to charm,
And flowers take no harm
Sprayed once each week with one in thirty-two.

We found this two-holer with one lid in our own village. The building, dated about 1650, is mainly Cotswold stone, to which the wooden front was recently added.

LAMENT FOR THE PRIVY

I'd like to go down to the privy again
Down the winding garden path
To sit again on that wooden seat
With no desperate need to rush.
With the old door halfway open
And the sun a-shining through
Back to the days of my childhood
When my world was green and new.

But what's the use of wishing
'Tis far too long ago
And the old wooden privies have rotted away
And the 'buckets' have vanished too.
Now we're up-to-date and modern
With a flush, and sometimes two,
But I still remember those happy days
When my world was green and new.

MOLLIE HARRIS

Right, when I was young, folk never spoke of stomach upsets or diarrhoea, but of having 'the back-door trots'. You can understand how this expression came about when you see just how close the back door of the cottage was to the family privy.

This thatched hut, with a door at each end, served two cottages at one time – there was not much privacy despite the partition separating the two privies.

When earth closets were destroyed on a farm during the mid-Sixties, this carved message was found on the wall of one of them. Now it is to be seen in the walled garden of the farm. Note the date: 1772.

A Privy by Any Other Name

A 'certain' place
Asterroom
Biffy
Bog
Boghouse
Bombay
Chamber of commerce
Chamberlain pianos ('bucket lav')
Chuggie
Closet
Comfort station
Crapphouse
Crapping castle
Crapping kennel
Dike
Dinkum-dunnies
Dunnekin
Dunnick
Dyke
Doneks
Dubs
Duffs
Dunnekin
Garden loo
Garder robe
Gong house
Gong
Go and have a Jimmy Riddle
Go and have a Tim Tit
Going to pick daisies
Going to see a man about a dog
Going to stack the tools
Going to the George
Going to the groves
Gone where the wind is always blowing
Heads
Here is are
Holy of holies
Honk
House of commons
House of office
Houses of parliament
Jakes
Jerry-come tumble
Jericho
Karzi
Klondike
Larties
Latrine
Lavatory
Little house
My aunts
Nessy
Netty
Out the back
Petty
Place of easement

Place of repose
Place of retirement
Reading room
Round-the-back
Shit-hole
Shittush
Shooting gallery
Shunkie
Slash house
The backhouse
The boggy at the Bottom
The bush
The dispensary
The dunny
The grot
The halting station Hoojy-boo (attributed to Dame Edith Evans)
The house where the emperor goes on foot
The hum
The jakers
The jampot
The japping
The John
The lats
The long drop
The opportunity
The ping-pong house
The proverbial
The Sammy
The shants
The shot-tower
The sociable
The tandem (a two-holer)
The thinking house
The throne room
The watteries
The wee house
The whajucallit
Three and more seaters
Thunder box
Two seaters
Widdlehouse
Windsor Castle
'Yer Tiz'

Especially for WCs:
Adam & Eve
Chain of events
Flushes and blushes
The penny house
The plumbing
The porcelain pony
The water box
Umtag (Russian version of the WC)
Going to inspect the plumbing
The urinal
Waterloo

The term 'privy' is an Early Middle English word which derives from the Latin 'privatus' meaning apart or secret.